series editor alexander payne

V.1 SERIAL BOOKS
DESIGN

1-901033-961

we like this!

michael young

The Serial Books Design series was conceived as a response to the need for a focussed and considered appreciation of contemporary designers and their work. The series intends to showcase the diversity of today's design, the new ways in which designers are reacting to materials and manufacturing processes and their responses to the wider culture within which their work is situated.

Where the late 80s and early 90s produced designers concerned with, amongst other things, 'eco' and salvage, what we are presently witnessing is something of how these designers are now shifting from small scale or one-off batch production to mass manufacturing. We are seeing the diversity of materials being used in their work and a new vocabulary being defined – from Marc Newson's 'Orgone' to the dual purpose furniture of the Azumi's to the reassessment and appropriation of everyday objects by Jam to Michael Young's renaissance of compound shape and form. Through the words, drawings and objects of the likes of the above designers Serial Books Design seeks to bring an awareness of their ideas and working methods to a wider audience, engendering debate while all the while celebrating the positive contributions being made by each of them.

I remember the launch of MY022 – a design complete in itself, the Magazine Sofa looking at me with its gaping, strangely animated mouth, Magnus the Basset hound as much a part of the image as Michael Young's sofa. Perhaps this was a further 'accessory' to create the new look for the 90s? – 'function' coupled with a laid back style of living, a by-product of the fashion, music and culture industries of the early part of the decade. The stark, white stand at 100% Design showcasing the MY022 collection – the pre-Muji metal box, complete with CD sized catalogue and T-shirt. There was undoubtedly something going on here. It appeared as though you, too, could become a part of MY022.

This was modernity with a fresh face, a modernity seemingly disregardful of what had passed before it. I heard later that what I had seen had been learnt on the shop floor of Space, Tom Dixon's studio – a space given to Michael's inspiration and searching for a particular purity and soul in and through design. Unusual here, and, again, something I was to find out later, that underneath all the styling and Japanese packaging – indebted to FLY's graphic design and the entrepreneurial skills of E&Y Co. – was Michael's very personal, even iconoclastic, attitude to design. This was an approach different to those of his peers.

After meeting Michael and talking through his work on several occasions I began to see connections to what had passed before him in the history of twentieth century design. I recognised his style and identity as being very particular reflections on our time. The way in which Michael worked – his cartoon-like sketches, poised to run off the page, on their way to the factory, to be produced after a signal back from Italy indicating "we like this!", marks a refreshing and curiously positive stance given where we find ourselves now – could things really be so simple? And his use of materials, which, if the technology were available, would be chameleon-like, ready to take on any colour to suit their surroundings or the tastes and whims of their owners – his temporary solution being the Smarty, 'dressed' and zipped-up in its neoprene skin, a pure form waiting for a new context – something akin, in approach, at least, to Gaetano Pesce's vacuum packed seating of 1969.

But Michael's fantastic imagination and child-like outlook on the world, in fact, sometimes provide a sober moment for reflection. Looking through the window of his "kitchen of the future" we see a treeless environment graciously provided with 'stick lights' so as to create an artificial Technicolor landscape with 'trees' reactive to the wind and, more intimately, to human touch.

This momentary duplicity aside, the key to any appreciation and enjoyment of Michael's work is to understand his simple desire in creating a surreal, fantastical world, a world that is "stress-free and relaxing". Up-tight could never be used to describe his designs, which are relatively unencumbered by preconceived ideas about culture, dreamed-up now, as he sits with a view of the Icelandic mountains, purposefully having removed himself from "cultural interference".

Michael Young's work, though, *is* in touch with its larger surroundings and can be seen as a reflection on these, be they those of the time of his living and working in London – the daze of the club scene of the early 90s – or his more recent, possibly more mature outlook, attuned, as it has become, to embrace the demands of mass production and manufacturing. Like past designers, such as Colombo, Panton and Mourgue, who responded, in their own ways, to the space age, Young has responded to the 90s in a non-provocative way, designing comfortable objects on which to sit, objects imbued with a purity and poetry. A precarious balance between an awkward humbleness and setting himself no boundaries, Michael's 'free-style' approach and way of working, I believe, has and will continue to inform his unpretentious designs.

Alexander Payne

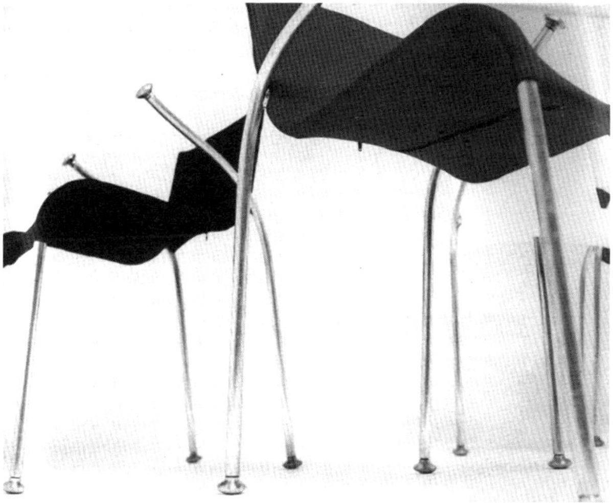

What I do know.

… if you look at the first thing I ever did, and, then, if you put everything else I did in a long row, there'd be this complete poetry running all the way through it – you'd be able to see, completely, the reasons for this first object existing. And within this first object, you can see this pattern…

Here is the poem I told you about. It was written in a Paris toilet by "Shel Silverstein". I think it's called "how many how much". I don't know if I even remember it correctly…

how many slams
in an old screen door
depends how loud
you slam it

how many slices
in a loaf of bread
depends how thin
you cut it

how much love
inside a friend
depends how much
you give them

how I used to work with Tom

I used to work from 10 to 5 with him, and then I used to 'play' in the workshop… I did this constantly, like day-in-day-out. I was covered in oil and burnt all over from welding the pylon furniture. I was just using his scraps and trying to make a living, and this was the only way I could figure out how to make work – just trying to make it and sell it, actually going to factories and seeing how they were doing things.
I think I really started thinking about organic forms when I started working with Tom.
I learnt a lot from him – about how he was using shapes… Tom had a very clear picture about how he wanted to push things.

I remember I was broke then, I literally couldn't afford to eat. All I would have during the day was a packet of crisps. It was pretty rough! But, naturally, the things that I learnt... I mean, Tom was using a definite logic, and though I sensed its influences, and the purity running through it, I was just looking in. I guess I realised from him that shapes were like a kind of logic.
Learning about this, which is something that a lot of people know – it's very obvious… but to a lot of people at college, they don't look around, so for them there is no meaning or understanding of what forms are. 'Looking around' is what I do because I am interested… which is something I learnt from Tom, although his work is really based on this concept of using modular shapes.

1 December 1993

To Whom it May Concern

This letter is written in support of Mr Michael Young's defence against a charge of Drinking and Driving which we understand will be heard in Court today.

We write to request the Court to consider the comments and information contained in this letter in respect of the above charge and how the court's decision, should it be adverse, may affect the current important stage in Mr Michael Young's career as a young and most gifted designer.

Mr Young is on the leading edge of young British designers. He has just this year, after considerable effort and personal hardship, been able to establish a firm business footing derived from his ability to design and most importantly, sell his first prototype designs. His lighting and seating designs have been bought by such leading retailers as The Conran Shop, (in both London and Paris), and as a result he will have his first One Man Show at an important gallery in paris next April, Galerie Mougin. Such a showcase for a young designer of his age is rare and nothing short of an accolade.

He is shortly to receive from the Crafts Council of Great Britain a Setting Up Grant to enable him to purchase his own necessary tools and equipment.

Mr Young commutes between London and the North East on a regular basis in order to co-ordinate the production of his work. Since he comes from Newcastle, he has chosen to give job opportunities to his home area in the production of the orders he is now regularly receiving. Because this production is now set up and running, he has to regularly drive between the North east and South east to transport materials and finished pieces. To this end his parents funded the purchase of a small van for him earlier this year and this has been instrumental in him being able to produce and sell his work.

In addition to his own work and to augment his income and therefore the funding of new prototype designs, we at Space have given him the full and free use of our workshop facilities, thus enabling him to develop prototypes and to manufacture small batch productions.

In return, he gives us invaluable driving and transportation assistance through the use of his own van for local London collections and deliveries. The contribution we make towards his mileage and petrol is a small but significant one, partly funding his own transportation costs.

We would therefore ask that you bear all of the above in mind when you consider the verdict to be made today as both Mr Young's career and livelihood are in the Court's hands.

Yours truly

I kept this letter through thick and thin. It explains a lot about how I was working at the time. A lot of work is as it is because of the access I have had to materials. I mean, I got so bored with exploiting a piece of MDF, of putting some foam on it and trying to make something of it. I really wanted to get more involved.

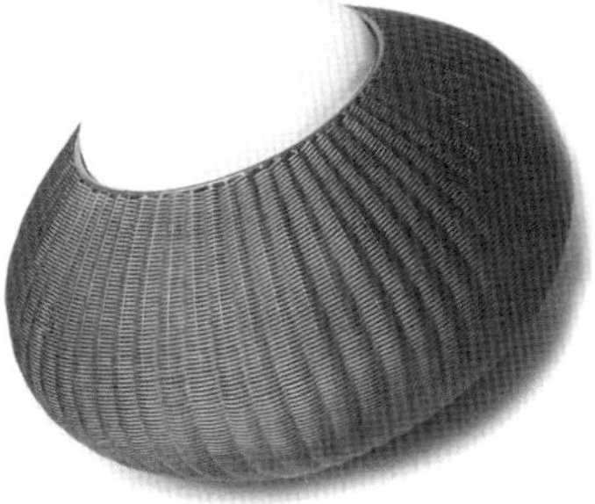

I do like shapes...

I had really just started learning about shapes. The way the Smarty was born was through my getting a huge block of foam… I put this massive metal bolt through the centre of it and attached it to the end of a lathe – the whole machine would be shaking up and down – I then cut at it with a circular saw blade, which was about a foot in diameter… I wrapped a cardboard box around my stomach, taped it to myself and cut into the foam with this really dangerous blade. Of course, I didn't know what it was going to be exactly… but it turned out to be one of the greatest accidents in my life. I remember Tom saying: "You've out minimalised Jasper Morrison!"

I am very much concerned with learning about 'form'. It's something I am passionate about. I feel that each time I do something new it's because I have learnt something from the last thing I did. I think this is what drove me to design… I realised that it was possible to create objects that I could happily provide a function. I always wanted to be 'creative', but my fundamental understanding was that to be this, I would have to be able to draw and I didn't enjoy or do this very well. So, 3D form was the perfect answer… That's not to say, though, that my work is just about form. But a chair generally needs four legs and a coffee pot needs a void so this "form follows function" bit has never been an entirely satisfactory lesson to me. The first book I read about all this was from the Bauhaus. I can't remember a thing about it now but I do remember the square, circle and triangle pictures… so I think the first things I designed came out of this in some way.

I have always been a lover of compound shapes because they expand your vocabulary of form, and thus your ability to resolve design problems in a more effective way. It's taken me a long time to find companies with processes that allow me to use compound shapes. They also take a lot of research as it is easy to make banal forms…

Much of my early work took its form because I only had access to certain pieces of equipment and I understood how to make what I wanted. Now that air moulding is available to me my work can be geared towards production. It's so easy for someone to criticise your work when they have no idea how clearly you have arrived at your design through your knowledge of a particular process. I think the first work I did with woven steel was an attempt in creating compound form but it was limiting because you can only work within a parallel framework.

I learnt a lot about metal from working at Tom Dixon's Space studio… also a lot about upholstering flat sheets of cut wood. I think sheet acrylic is quite a nasty material, I don't think it has much going for it so I have only used it once. Perhaps such simple processes force you to go deeper into the poetry of your design. I think so, at least. It's a good foundation for me to use new shapes. My new work with MAGIS should reflect this.

I have a basic philosophy – each piece has an organic form, a hidden vocabulary...

These are really more advanced studies of form, and sometimes I don't think of their function, because I am, here, really just manipulating the shapes and making them into objects.

I like to produce things in white because I want the design to be represented very clearly. Colour can redefine forms. If I could find a simple effective way for the customer to apply the colour when they got home I would rather do that because I wouldn't know and they would be happier.

When I designed the Smarty it was done with the intention that it could be produced in any colour. It's the nature of the form that's important – there's no reason to restrict it.

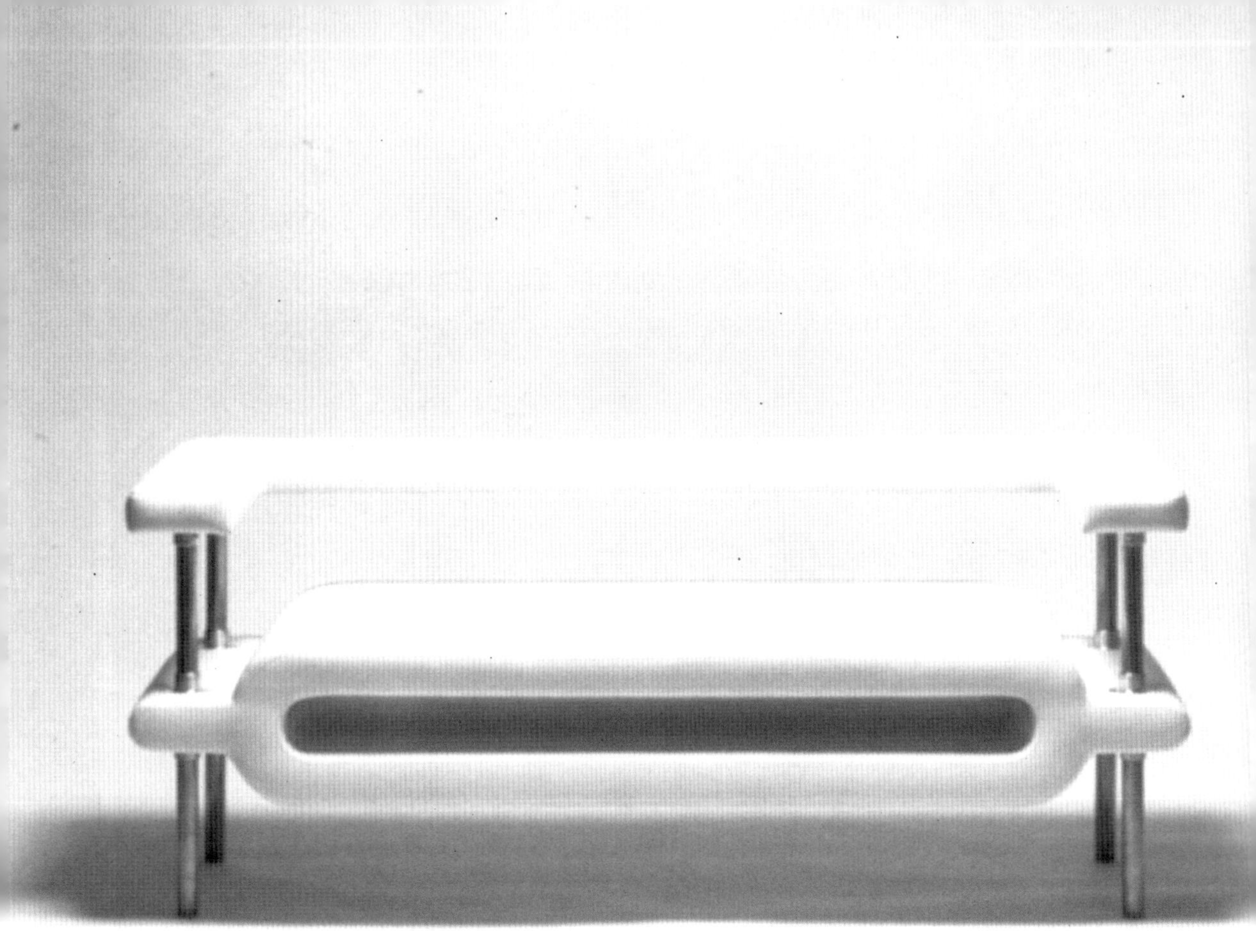

this is strange,
the link it has
is that I really
like this scarf...

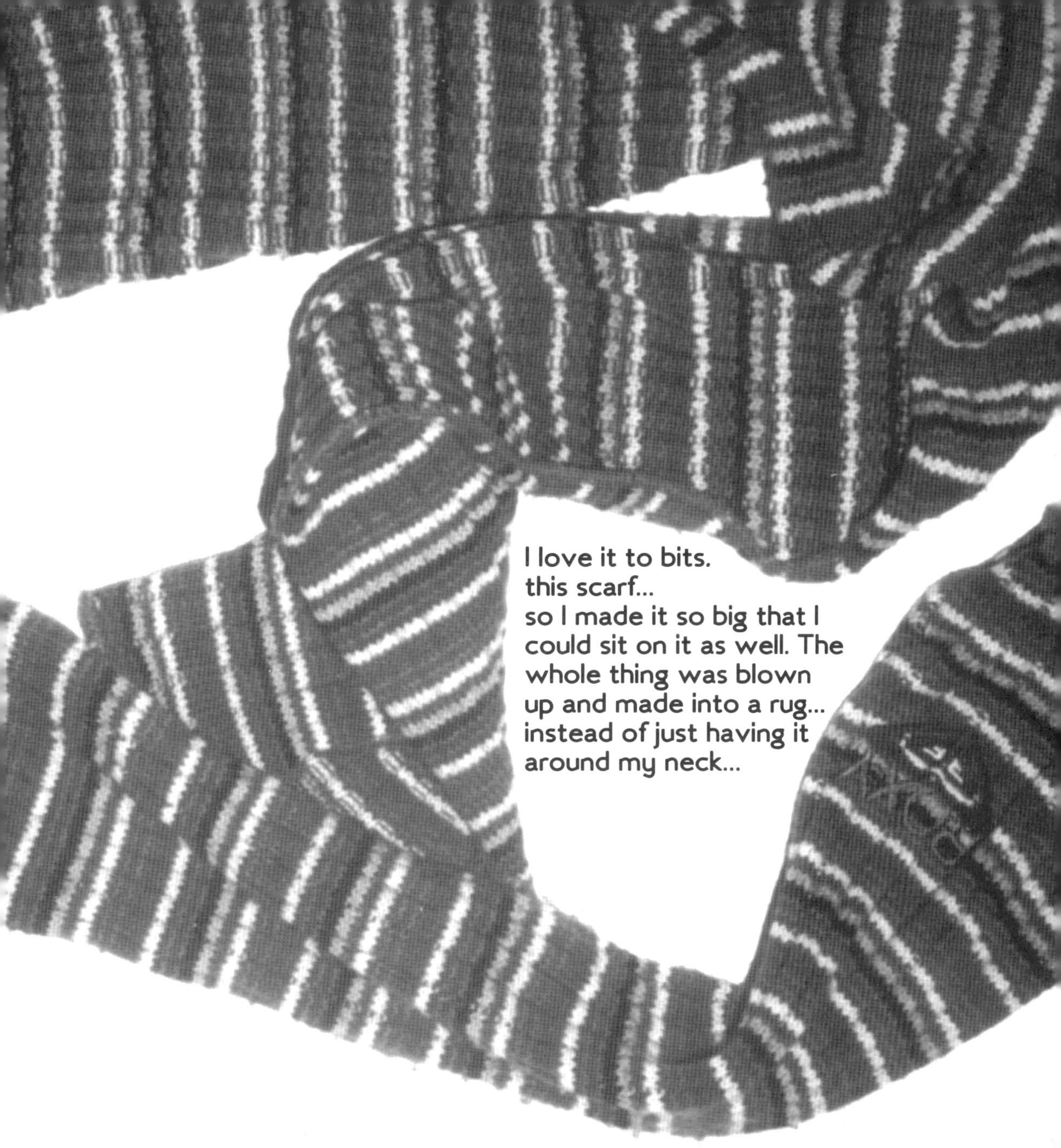

I love it to bits.
this scarf...
so I made it so big that I could sit on it as well. The whole thing was blown up and made into a rug... instead of just having it around my neck...

texture...

size... fragility...

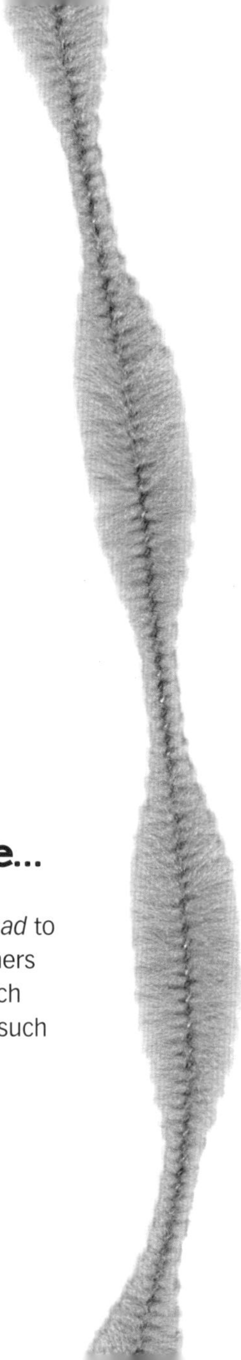

I love texture...

it's really important. I *had* to collect these pipecleaners from Japan. They're such eccentric objects, and such beautiful colours...

I love the interplay of obscuring the scale of things... of having no definitive boundaries around what I'm doing.

I like the fragility of objects as well. like Jeff Koons' "balloon dog" because it's a mixture of all of these things. I feel an empathy with conceptual art. Taking these interests into account, I think it should be clear that I am not inspired by 60s design or even retro. It is unfortunate that one misguided journalist can influence another. The textile on the Slit Sofa is as close as I have come to borrowing something as I have found it – from a Japanese CD cover… I fell in love with this pattern so I printed the fabric on to this design to create something that almost comes out of the cover, out of the computer, like a toy… a big toy.

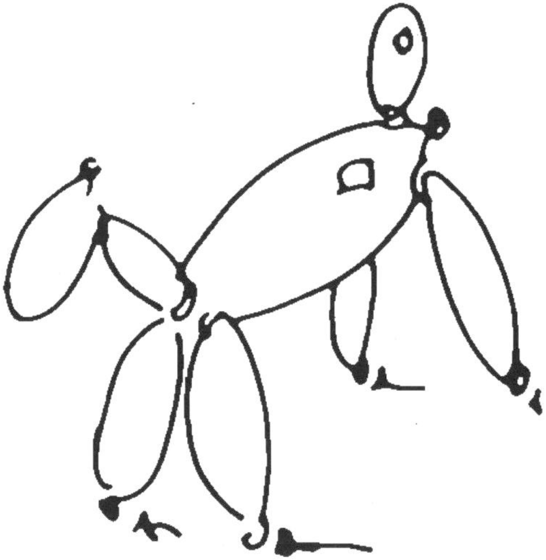

I've not been directly influenced by Japanese things, their purities and forms, etc., although in a certain context I feel very spiritually connected to the essence of things… I feel something. I did a show in an old tea house in Kyoto, there was a very magical air to it… a very natural joining of environments. I think that good design is created through integrity… a genuine soul. So design that is created with these qualities should assist in creating the soul in a room. When you feel uncomfortable in a room its because of a lack of soul. The Japanese are, in a traditional sense, very good at creating atmosphere through design. It's purely a personal preference for comfort which has influenced bringing objects closer to the floor, it's not because I am mad about Japan. Mind you, I do find legs a bit of a struggle, they're often just a necessary protrusion… or intrusion.

I've often been referred to as someone doing minimalist work,

except that I say "fuck off" to everything a bit more. I just feel very spiritual about what I am doing because I'm being myself, there's no bullshit, just my thoughts...
I suppose that's the case with most people really. It's more productive to be yourself.

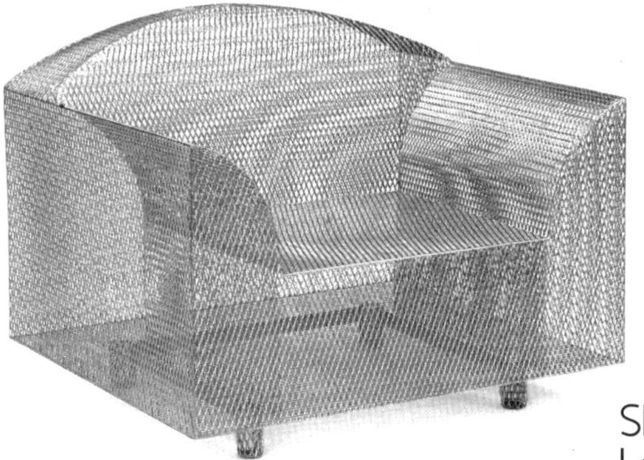

Shiro Kuramata...
I enjoy the Zen of his work so much,
the purity, I love it.

I have an affinity with things that are very straightforward and understandable mechanically. I like milk bottles because their function is so clear, the design has changed very little as far back as the Roman era. They also have a 'surrealness' because the milk is very white and comes from cows which are very strange animals. I find the Tetra Pak completely charmless compared to this, it's an unflavoured design solution, that actually doesn't work very well.

COKE
CO
OP
Coca-Cola

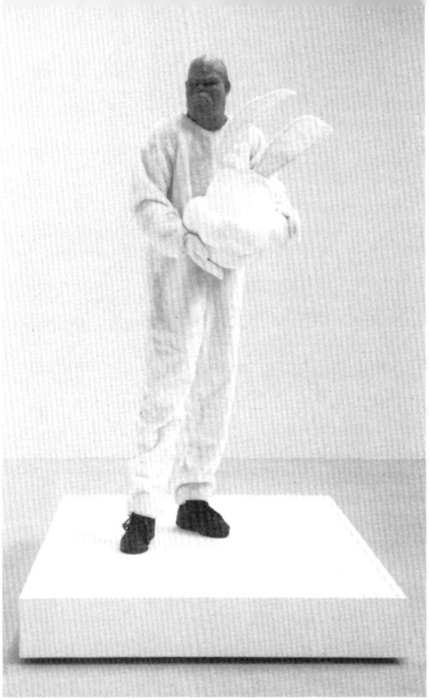

I am often quoted as a Britpop designer

and all that stuff, but I've never felt a part of that scene. Britpop is more relevant to music and art, but I've not paid much attention to it. I've always liked Nick Cave and Motorhead, which I don't think as being street level. I can't handle most magazines either, except for, maybe, Frieze and Blueprint.

I have always been a great fan of John Issacs' work – he is one of my favourite sculptors.
I like the concept of it a lot...
I like the tragedy.

It's just a fairytale, a sort of Fantasia, that's what I like about it – I love the sculptures and things.

It's very difficult to put it into words, what effect Edward Scissorhands has on me, but it's just the whole vibe, basically, of fantasy, a complete, nice, great little world – I want to bring that kind of sensation to life. And that, ideally, is the challenge… Originally, when I left college I was going to create these worlds, an existence at night, when you're sitting at home – I used to have this kind of fantasy which feels great to be in, like a game or something.
It is important because it's a magical and melancholic fairytale. I think it's a great story because it creates poetry and provides answers. It has a dark humour about it, the way it looks so light-heartedly at the effects of consumerism on society, the greed, the love and the hate…

Edward Scissorhands

I was asked by CASA BRUTUS, in Japan, to design a "kitchen of the future", so I thought I'd just change the people... and from that a whole load of new opportunities came my way.

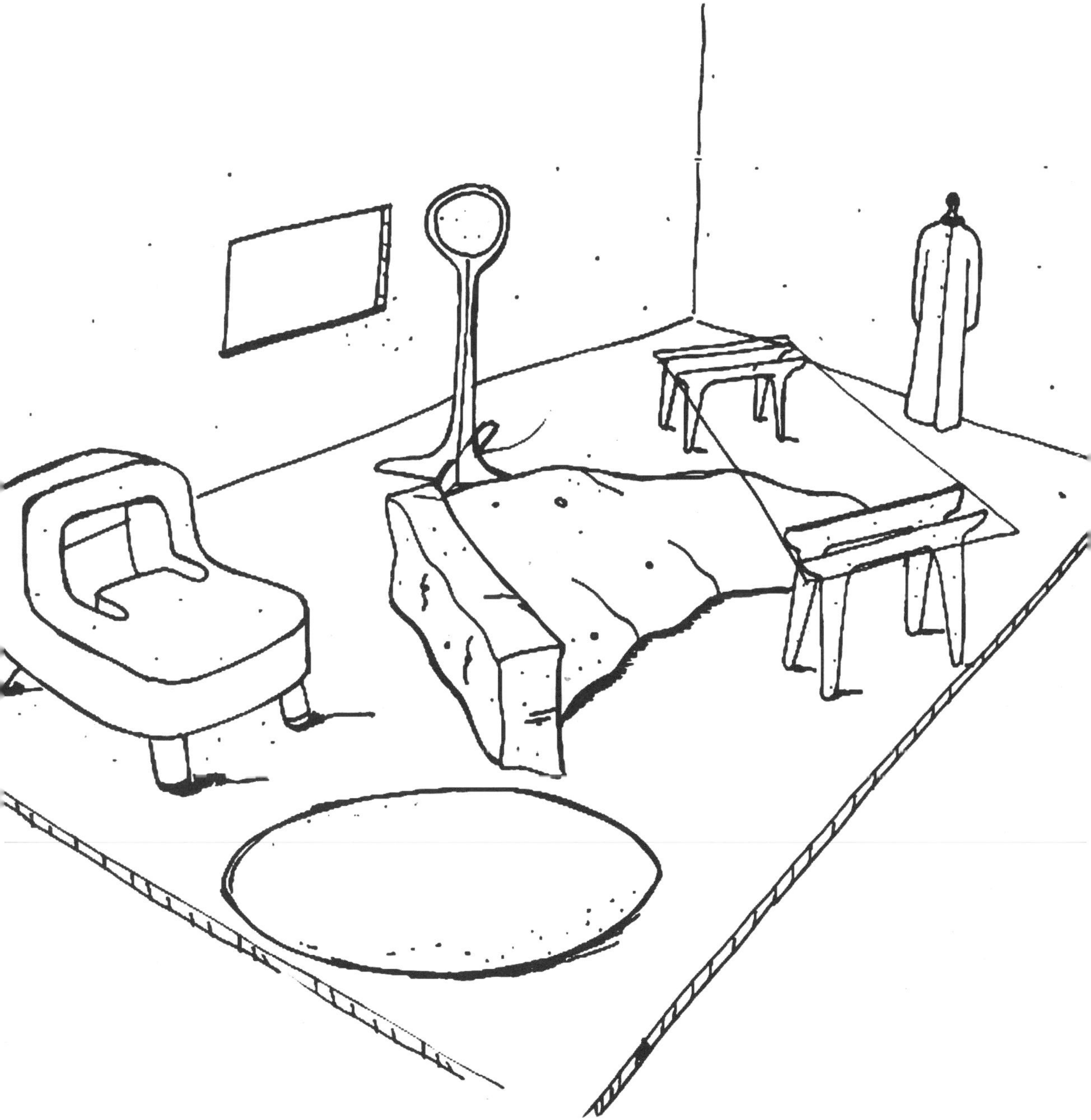

ingredients:
puffinbreast, water,
tomatoes,
vegetable oil,
cream,
milk protein,
flavour enhancer,
herbs,
spices.
nutritional facts
per 100g
energy 189kj/45kcal
protein 0.5g
carbohydrate 4.8g
fat 1.8g
295g
ffinsoupp
nsouppuff
with tomato with tomato
toith tomato with t

young&beautiful's
homemade
soup puffinsoup puffins
with tomato with tomato with tomato with

at home I can see mountains,

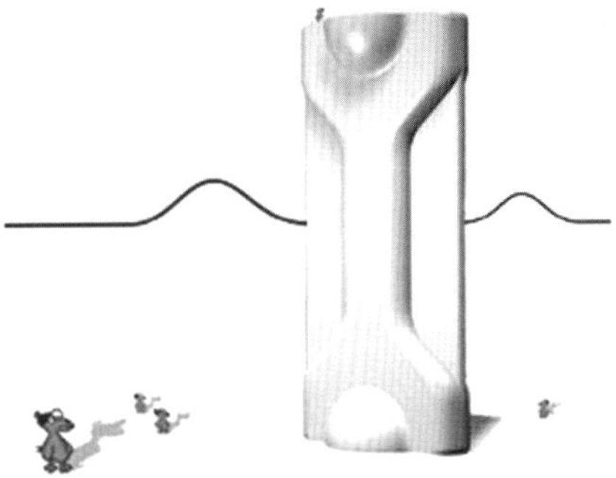

icy mountains outside, which is fantastic. I was going to move out to the countryside, but then I thought, "this could get a bit weird! I could be classified as a recluse!" so I'm going to move right bang in the centre of this place.

... I have my own vocabulary that I like to evolve. I don't want the interference of consumerism or culture.

But I suppose there must be some interference... ingredients, perhaps, provided by the early worlds of Tom Dixon and Ron Arad. I mean by this their attitude as mavericks, because the abrasiveness of their early work never appealed to me.

It is very important
for me to have this
in the room.

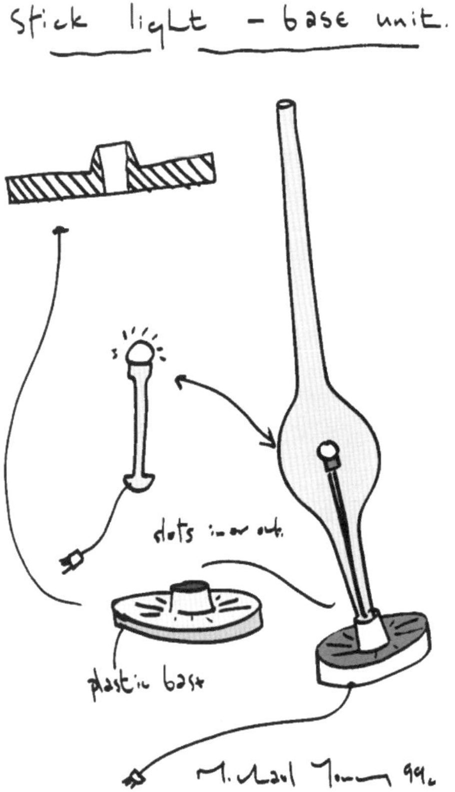

I hope that in the future, once we know more about DNA, we will be able to recreate molecular structures out of some base material, and so be able to remake any type of material, Then we can really put something back into the environment. The column which eurolounge are now producing in a simpler format came from the garden in the "kitchen of the future", the idea being that there was no oxygen left so the trees were recreated electronically and the wind would make these little lights change colour – to replace the flora. I made one such kinetic tree for an exhibition in Japan, with the help of Jeremy Lord and immense input from musician Kenjiro Matsuo, who raised about £50,000.00 for it. We went on to show it in a technically reworked format at the Paris Salon, then in Cologne and finally at Milan in 1998, with Cappellini. Anyway, the next step is to complete the kitchen window idea with a show on a volcanic landscape in Iceland…

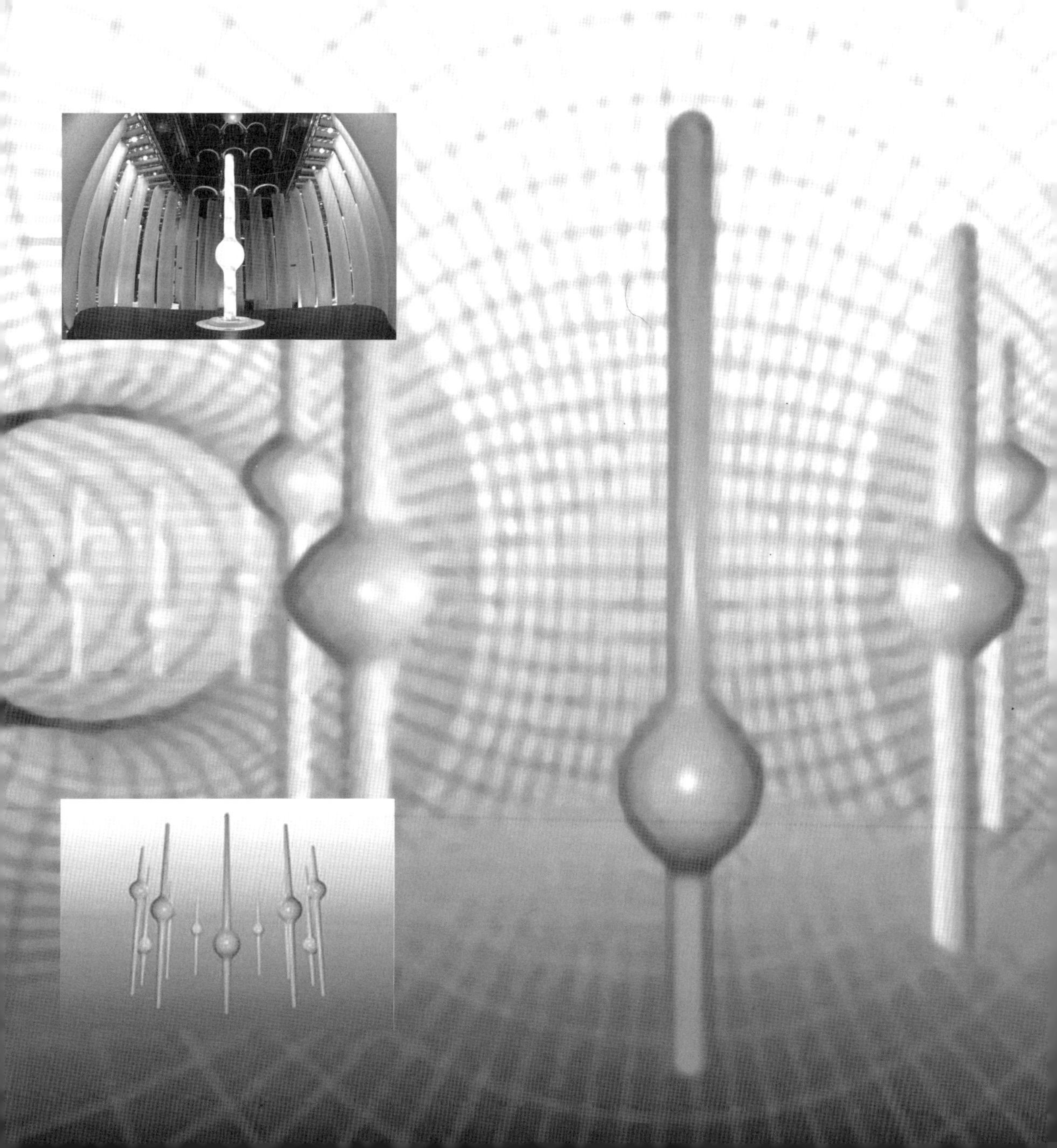

When I got married in Iceland I had to design a ring... so this ring, made in white gold, was the obvious thing for me to do. The company that made them for me then asked if I would set up a new division for them... which I did with my wife. We called it S.M.A.K. Iceland. Consequently there are designs from Ronan & Erwin Boulourec, Michael Marriott, Inke Hans and Matalie Crassett... isn't it nice how things are born!

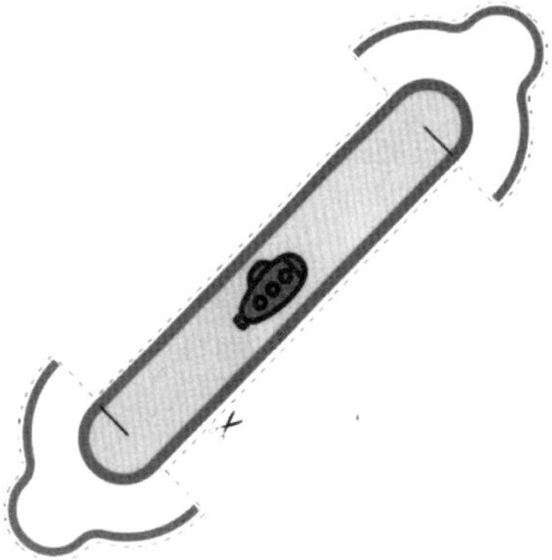

I like the object's construction and function to be clear, though I find objects that are deliberately minimal in the fashionable sense inhuman and unusable. I believe in creating warmth and a human relationship or quality in objects. If you play with forms there is always the opportunity of a function arising… Funnily, I only draw now as I don't get a chance to work in other ways. I have used a block of styrofoam on board an aircraft recently, but was requested to 'cease' my activity.

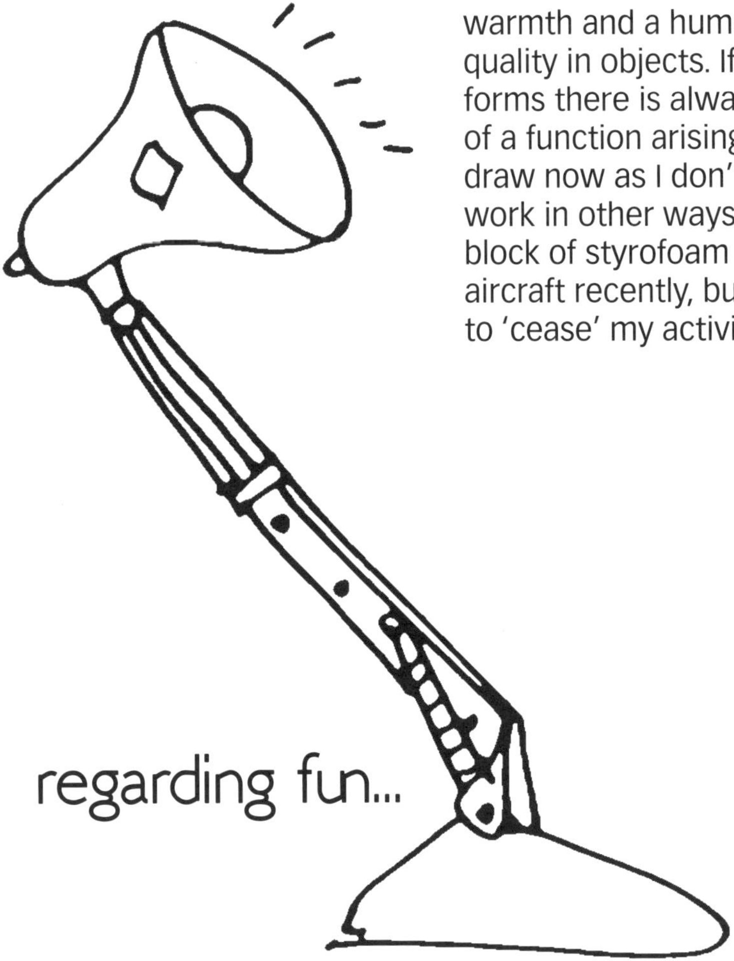

regarding fun...

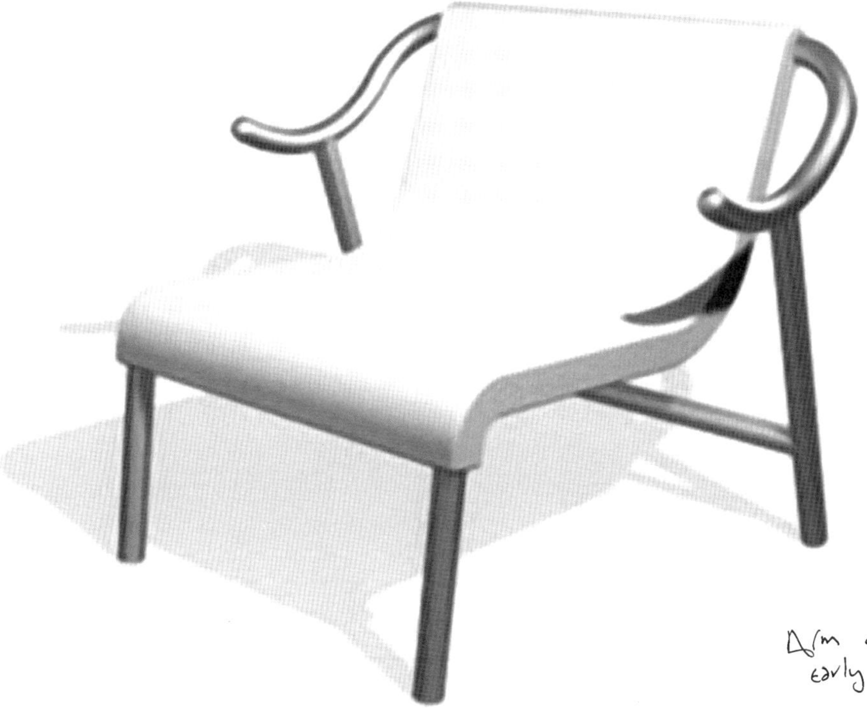

Arm chair —
early concept to develop

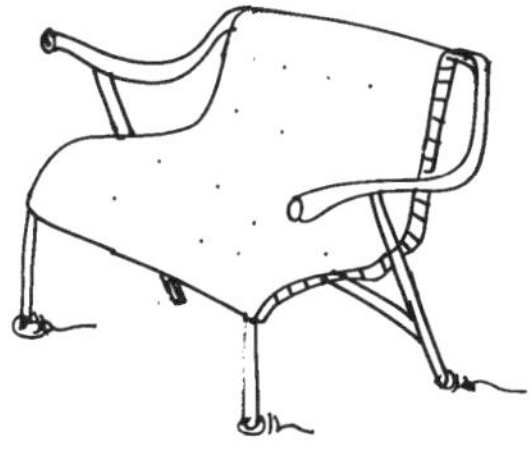

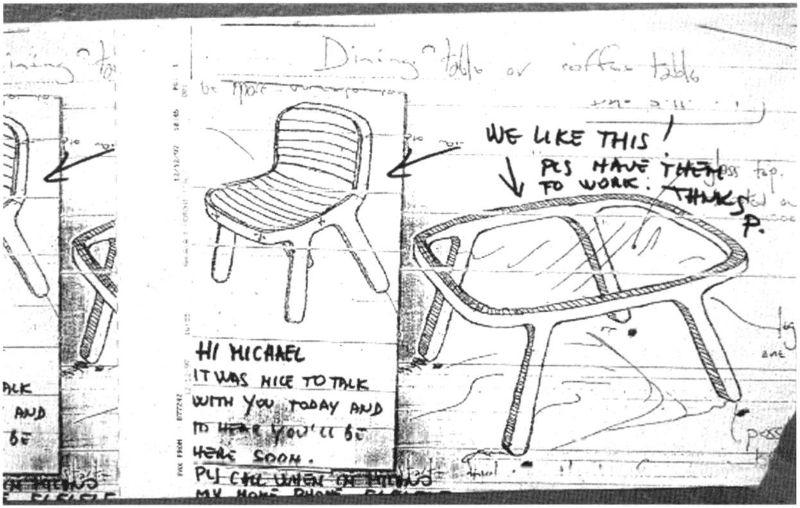
Dining table or coffee table
WE LIKE THIS!
PLS HAVE THEM TO WORK. THANKS P.
HI MICHAEL
IT WAS NICE TO TALK
WITH YOU TODAY AND
TO HEAR YOU'LL BE
HERE SOON.

I've never been inspired by looking at design,

although I appreciate good
design. It doesn't really
influence me, if it did I would
have given up by now.
I believe things must come
from within.
I think I live in my own
inspirational kinetic world…
I mean, if you are doing
five projects at once you
create new variants…
like in genetics…

Iceland is a great place to find old tractors and spend time thinking about why you like them… I still don't know what draws me too them in any great detail. Perhaps this has to do with an aesthetic being expressed through it's function. They almost look like big mechanical components. They look so unusual because of the level of technology they use. This is their charm. I don't like old cars in the same way because the technology used in them at this time was used also for styling.

"Our imagination and our dreams, are forever invading our memories; and since we are all apt to believe in the reality of our fantasies, we end up transforming our lies into truths. Of course, fantasy and reality are equally felt, so their confusion is a matter of only relative importance."

my last breath
Bunuel

Magazine Ring
together with
Katrin Petursdottir
sterling silver or gold
S.M.A.K. Iceland
1998

MY 08
Occasional Shelf
wood and tubular
aluminium
280mm x 280mm x
1670mm high
MY 022 UK Ltd
1995

MY 15
Bar stool
spun aluminium and
neoprene or vinyl
440mm x 30mm x
800mm
MY 022 UK Ltd
1995

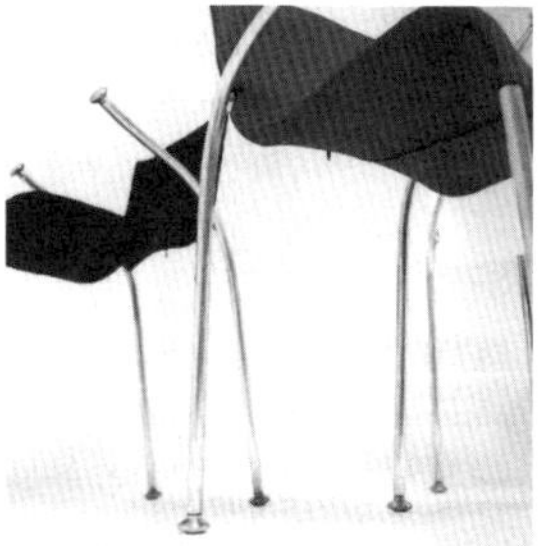

MY 01
Cafe Chair
tubular aluminium and
neoprene
800mm Ø
MY 022 UK Ltd
1995

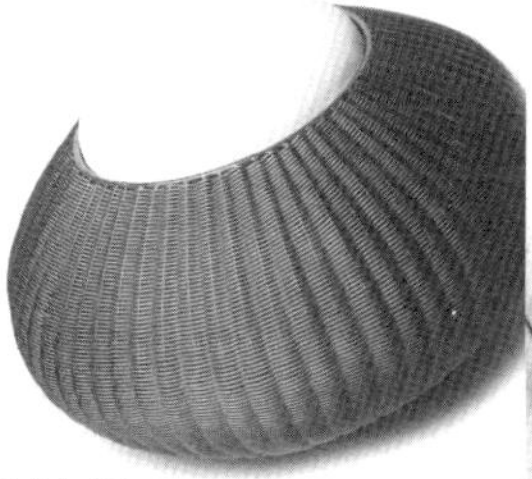

MY 17
Putz
stainless steel cable
and neoprene
600mm Ø x 560mm
high
MY 022 UK Ltd
1995

MY 16
Smarty
pu and neoprene
800mm Ø
MY 022 UK Ltd
1995 2nd edition
re-issued by Cappellini
1998

MY 16
Smarty
turned foam and flock
800mm Ø
MY 022 UK Ltd
1994 1st edition

MY 13
Woven Light
stainless steel cable
and nickel
110mm x 230mm high
MY 022 UK Ltd
1995

MY 07
Magazine Coffee Table
pu and aluminium
480mm x 350mm x
470mm high
E&Y Co. Japan/
MY022 UK
1995

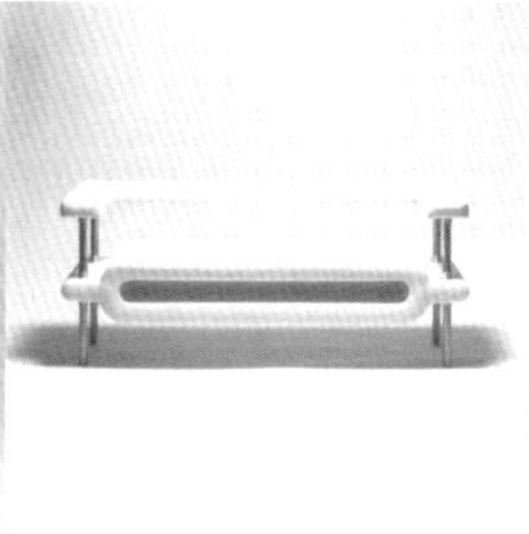

MY 03
Magazine Sofa
vinyl aluminium
620mm x 1420mm x
640mm high
E&Y Co. Japan/
MY022 UK
1995

Profiled Honeycomb
Screen
aluminium
one-off
1995

MY Scarf Carpet
wool
2500mm x 1800m
Christopher Farr
1998

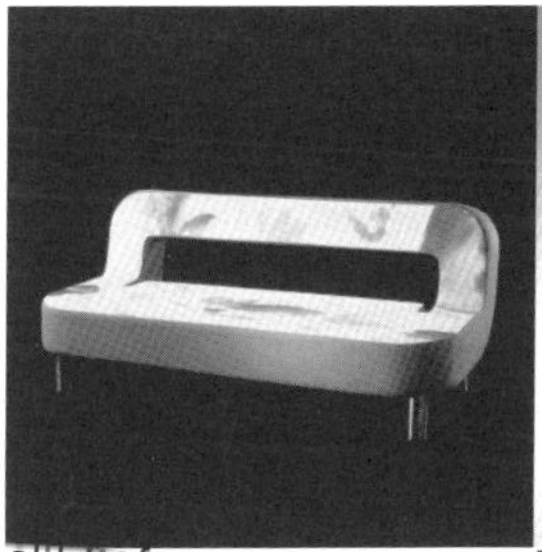

Slit Sofa
printed fabric and aluminium
1000mm x 850mm x 750mm
Concept for TT, made by Domeau Peres
1998

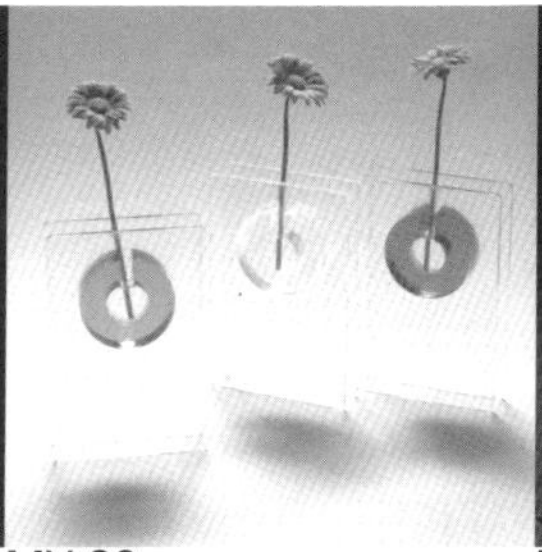

MY 30
Flower Vase
acrylic
150mm x 45mm x 230mm
MY 022 UK Ltd
1996

MY 27
Fly Sofa
single, two and three seaters
white vinyl aluminium and wood
1970, 1550, 1200mm x 840mm x 570mm
E&Y Co. Japan
1996

MY 69 Vase
sterling silver
110mm x 280mm
Sawaya & Moroni
1998

Kitchen of the Future
cartoon illustration for
brutus magazine
1996

Matrix of Amnesia –
Cappellini
together with
Katrin Petursdottir and
John Issacs
London
1998

Home-made Puffin
Soup Label
a Young & Beautiful
product
1999

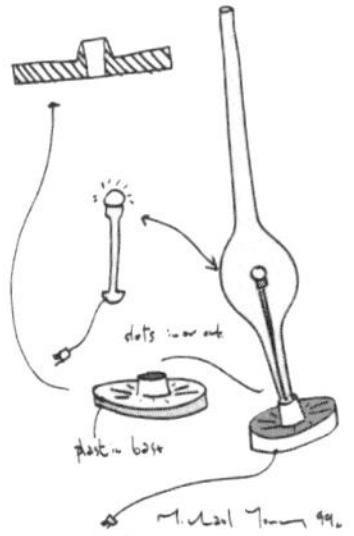

Light Stick
polypropylene
1700mm high
eurolounge
1999

Synapse Column
together with Kenjiro,
FLY, Jeremy Lord and
Inflate
fibreglass and electrical
components
3200mm high
MY 022 UK Ltd one-off
1997

Magazine Necklace
together with
Katrin Petursdottir
sterling silver or gold
S.M.A.K. Iceland
1998

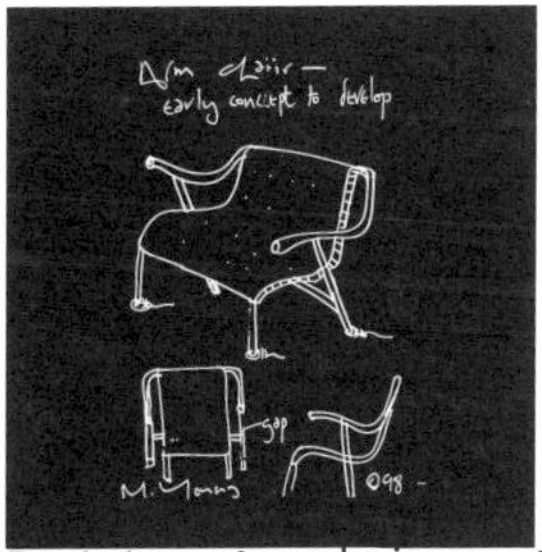

Prototype Armchair
launch date 2000

MY 068
Wood Chair
natural or stained birch
Sawaya & Moroni, Italy
1998

Bibliography

Born
1966 Sunderland England

1989-92
Attends BA furniture & product design course at Kingston University in London whilst simultaneously working at Tom Dixon's studio Space

1992-94
Works at Space and starts developing first collection with the aid of British Crafts Council "setting up" grant. Shows first collection of woven steel furniture in group shows curated by E&Y Co. in Hong Kong and Tokyo. Shows in Works of '94, London with Michael Marriott and Jane Atfield. First one-man show in Paris with Gladys Mougin and launches Smarty.

1995
Sets up first studio in London to work on exhibition and collection for E&Y Co. Tokyo. Magazine Sofa launched. Wins talente 95 in Germany with profiled honeycomb screen. Sets up MY-022 concept/ production studio in London. Projects with SOM architects.

1996
Conceptual installations in Kyoto, collection purchased by Neue Sammlung Museum Germany, Design Museum, London, Musée de Arts Décoratifs, Paris and Municipal Museum of Portugal. Launches Fly Collection. Shows in Neotu, Paris. Furniture for boutiques in Japan.
Carpets for Christopher Farr.

1997
Designs Synapse column to celebrate 50 years of Japanese light railway in Fukuoka. Concept travels to Paris, Cologne, London and Milan. Collaboration with Cappellini, Italy. Ceases MY022 production to concentrate on project work. Works with Englender, UK.

1998
Smarty remanufactured by Cappellini. MY068 chair made by Sawaya & Moroni, Italy. Relocates to Iceland, collaboration begins with Magis, Italy and Rosenthal, Germany. Jewellery collection for S.M.A.K. with wife Katrin Petursdottir. Works on production version of Synapse column. Furniture for "the cigar club". Carpets for Christopher Farr.

1999
Founds "Young and Beautiful" concept company. New product ranges for Magis, Rosenthal, Cappellini, Ritzenhoff, Englender and eurolounge

Selected Exhibitions

1994
Touring Show, E&Y Co.,
Tokyo and Hong Kong
Group Show, Crafts Council,
London
Galerie Gladys Mougin, Paris

1995
E&Y Co. Tokyo
talente 95, Munich

1996
Neotu, Paris
100% Design, London

1997
Salon de Mobile, Paris
Internos, Milan
Synapse, Fukuoka
Schmit, Konstanz
Synapse Installation
Space, London

1998
Group Show
Galerie Artificial, Cologne,
Presentation
Englender, London
Presentation
Christopher Farr, London,
Presentation
Sawaya & Moroni, Milan,
Presentation
Cappellini, Milan
Group Show
Totem, New York
Matrix of Amnesia
Cappellini, London,
Group Show
Oikos, Athens

1999
Group Show
Galerie Artificial, Cologne
Presentation
Salon de Mobile, Paris,
Group Show
Kjarvalsstaoir Museum,
Reykjavik
Concept Solo
Orzzonti, Tokyo
Concept Solo
JA, Los Angeles
Concept Solo
Moss, New York
Light Stick
eurolounge, London

Work in Public Collections

Design Museum, London
Musée de Arts Décoratifs, Paris
Neue Sammlung Museum,
Germany
Municipal Museum, Portugal

British Library cataloguing-in-publication data.
A catalogue record is available from the British Library.

Serial Books Design was conceived, edited and produced by Duncan McCorquodale

Designed by
Maria Beddoes &
Paul Khera

Printed in the
European Union

ISBN 1 901033 961

Thanks to
Sarah Allen
eurolounge
Paul Hughes Fine Art
John Isaacs
PNP/Patrizia Mezzanzanica
David Simmons

Black Dog
Publishing Limited
PO Box 3082
London NW1 UK

Forthcoming
Azumi
Tom Heatherwick
Jam
Marc Newson

For further information
concerning other titles in the
Serial Books Design and
Serial Books Architecture &
Urbanism series'
please contact Black Dog
Publishing Limited in writing.

Architecture Art Design Fashion Photography Theory and Things